WHAT SPRINGS OF RAIN

Flora and Fauna of the Amazon Rainforest and Cloudforest

Lindsay Erin Lough

SUNBURY PRESS

Mechanicsburg, PA USA

Published by Sunbury Press, Inc.
Mechanicsburg, Pennsylvania

SUNBURY
P R E S S

www.sunburypress.com

For information about special discounts for bulk purchases, please contact Sunbury Press Orders Dept. at (855) 338-8359 or orders@sunburypress.com.

To request one of our authors for speaking engagements or book signings, please contact Sunbury Press Publicity Dept. at publicity@sunburypress.com.

ISBN: 978-1-934597-52-1 (Hardcover)

Library of Congress Control Number: Application in Process

SECOND SUNBURY PRESS EDITION: May 2019

Product of the United States of America
0 1 1 2 3 5 8 13 21 34 55

Designed by Crystal Devine
Cover by Lawrence Knorr
Cover photo by Lindsay Erin Lough
Edited by Lawrence Knorr

Continue the Enlightenment!

Dedicated to the memory of Pablo Barba, who passed away during his courageous research on the Amazon rainforest.

FOREWORD

December 2007
Near Los Amigos Biological Station
in Peru

Morning mist. Quietude. Scattered shafts of sunlight slant down through random holes in the jungle canopy. Bright green splotches shine where the light strikes the murky forest floor.

We walk rapidly, but softly, along a narrow overgrown path in the rising heat and humidity of the upper Amazon rainforest. My daughter, Erin, leads the way with her machete in hand and her Nikon telephoto draped securely around her neck. She wears a gray sleeveless tank top, her short blond ponytail tied with the back of her red forehead bandana. Her protective, knee-high, thick rubber jungle boots swish against the thick vegetation. Her skin glistens with sweat as we work our way along the Madre de Dios River in Peru. She moves comfortably in the jungle.

We follow normal jungle hiking protocol by carefully watching the path just ahead of each step to avoid treading on anything that might prove dangerous. Bites, stings, scratches, rashes, and puncture wounds accrue to those who are careless with their hands and feet in the jungle.

Sharp, snapping, rustling sounds in the trees forty meters ahead break the pervasive stillness. We glance up to see the troop of saddleback tamarins we have been following. We continue approaching, hoping to get close enough for an exceptional photo op. Looking up and down from path to monkeys to path again, while trying to walk steadily, is not, for me, as easy as it sounds.

I see Erin loose her camera and slow her pace. Just then, she steps across a yellow, red, and black banded rope that moves across the path and disappears even as I have to jump to avoid stepping on it. I call a warning to Erin, who turns, and with me abreast, cautiously approaches the spot where the snake disappeared. Gingerly, Erin pushes aside the low growth with her machete. And there, again, is the snake. After a hesitation, the small black head turns away and the colored bands move quickly, farther into the concealing vegetation. We whisper excitedly to each other, but decide not to follow the snake into a wide gully along the path where the brush is taller and thicker, making visibility very poor and increasing the risk of a surprise attack. Reluctantly, we leave the snake to take up our pursuit of the saddlebacks. Later, we regret we didn't get a photo.

That afternoon at the research station we confirm, from reference book descriptions and pictures, our encounter with a deadly coral snake. We laugh giddily, we embellish the tale, we engage in some braggadocio and we hug each other in the wild, nervous joy of remembering our experience.

See pages 19 through 21 for pictures of saddleback tamarins.

WEL
July 2018

A granular tree frog (Hypsiboas cinerascens) sits defiant on its leaf. Granular tree frogs are both arboreal and nocturnal, although males will sometimes call for mates from hidden leaves near the surface of a stream or standing water. April 2008. Los Amigos Biological Station, Peru.

KERMIT

PEEKING

A veined tree frog (Trachycephalus venulosus) peeping out of a bamboo stock. Adult veined tree frogs secrete a sticky, white, poisonous, alkaline mucus from their skin and must be handled carefully. They are also notoriously difficult to catch because adults can parachute from trees to safely glide as far as 27 meters away. March 2008. Los Amigos Biological Station, Peru.

A slender-legged tree frog (Osteocephalus sp.) that was just old enough to venture our of its watery home as a young tadpole. However, it still carried the tell-tale sign of its life before metamorphosis: a long tail and barely visible remnants of gills. Newly transformed frogs typically keep their tales for at least a few days after they have left the water. April 2008. Los Amigos Biological Station, Peru.

METAMORPHOSIS

Flora and Fauna of the Amazon Rainforest and Cloudforest

LIZARD LIPS

A collared tree lizard (Plica plica), also known as the South American chameleon, is commonly found on trunks of Amazonian trees. Anthropologists say the thrashing tail of the collared tree lizard is a recurring motif in ancient Amazonian petroglyphs as a phallic symbol. April 2008. Los Amigos Biological Station, Peru.

This young brown-banded water snake (Helicops angulatus) was found in a brackish stream near the Los Amigos Biological Station. It was about the size of a pinky finger. Brown-banded water snakes are nocturnal, aquatic snakes that sometimes have the rare characteristic of bearing viviparous young— or giving birth to live young. March 2008. Los Amigos Biological Station, Peru.

TIMID

UNLIKELY CAMOUFLAGE

A uniquely adapted pink grasshopper blending into the flowers of a touch-me-not plant (Mimosa dormilona). June 2008. Los Amigos Biological Station, Peru.

Flora and Fauna of the Amazon Rainforest and Cloudforest

A slug moth caterpillar from the family Limacodidae. These caterpillars glide on liquid silk and suckers rather than walking with legs like most other caterpillars. March 2008. Los Amigos Biological Station, Peru.

ALICE'S CATERPILLAR

Flora and Fauna of the Amazon Rainforest and Cloudforest

CATERPILLAR MASS

A mass of caterpillars of the Saturniidae family clustered on a tree trunk. They will eventually metamorphosize into regal or giant silk moths. February 2008. Los Amigos Biological Station, Peru.

A butterfly competes with a few ants for the sweet contents of this piece of honeycomb. May 2008. Los Amigos Biological Station, Peru.

SIP HONEYCOMB

Flora and Fauna of the Amazon Rainforest and Cloudforest

ALIGHT

A group of butterflies gather on a bare hanging branch to rest for the night. April 2008. Los Amigos Biological Station, Peru.

A sapphire-colored dragonfly rests on an overhanging branch of an Amazonian palm swamp. June 2008. Los Amigos Biological Station, Peru.

ARC

SEA ANEMONE

White mushrooms clustered on a log mimic ocean fauna. June 2008. Los Amigos Biological Station, Peru.

Two rare and colorful Amazonian orchids. Los Amigos Biological Station, Peru.

IRIDESCENT ORCHID

FIRE NEBULA

Flora and Fauna of the Amazon Rainforest and Cloudforest

BEAUTY IN THE PALM SWAMP

In the midst of a palm swamp filled with anacondas, a rare aquatic orchid emerges. May 2008. Los Amigos Biological Station, Peru.

A rare Amazonian canopy flower. December 2007. Los Amigos Biological Station, Peru.

FIREBURST

Flora and Fauna of the Amazon Rainforest and Cloudforest

SUNSET TREE

A sunlit canopy tree as seen from a 60-meter tall tower. June 2008. Los Amigos Biological Station, Peru.

A brown capuchin monkey (Cebus apella) defending its territory. June 2008. Los Amigos Biological Station, Peru.

TERRITORIAL

WHAT SPRINGS OF RAIN

Flora and Fauna of the Amazon Rainforest and Cloudforest

OVERLOOKING THE KINGDOM

A rarely seen wild emperor tamarin (*Saguinus imperator*) taking a moment to scan his forest domain. Emperor tamarins wiggle their distinctive white mustaches to communicate with one another, most commonly to signal aggression or warning. November 2007. Los Amigos Biological Station, Peru.

A group of saddleback tamarins (*Saguinus fuscicollis*) feasting on a tasty annona tree fruit. May 2008. Los Amigos Biological Station, Peru.

BREAKFAST ANNONA

Flora and Fauna of the Amazon Rainforest and Cloudforest

FIRST ASCENT

A young saddleback tamarin (Saguinus fuscicollis) on a venture away from its mother's back. May 2008. Los Amigos Biological Station, Peru.

A young saddleback tamarin (Saguinus fuscicollis) had become lost from its family group and was calling for help. February 2008. Los Amigos Biological Station, Peru.

ABANDONED

Flora and Fauna of the Amazon Rainforest and Cloudforest

CAPTURE A marsupial (Marmosops sp.) was captured for research purposes at the Los Amigos Biological Station, Peru. September 2007.

A fruit bat (Artibeus sp.) flying free after researchers gently studied and weighed it. April 2008. Los Amigos Biological Station, Peru.

FREEDOM

Flora and Fauna of the Amazon Rainforest and Cloudforest

ORNATE HAWK-EAGLE

A rarely seen ornate hawk-eagle (Spizaetus ornatus) holding its prey: a large wood quail from the forest floor. October 2007. Los Amigos Biological Station, Peru.

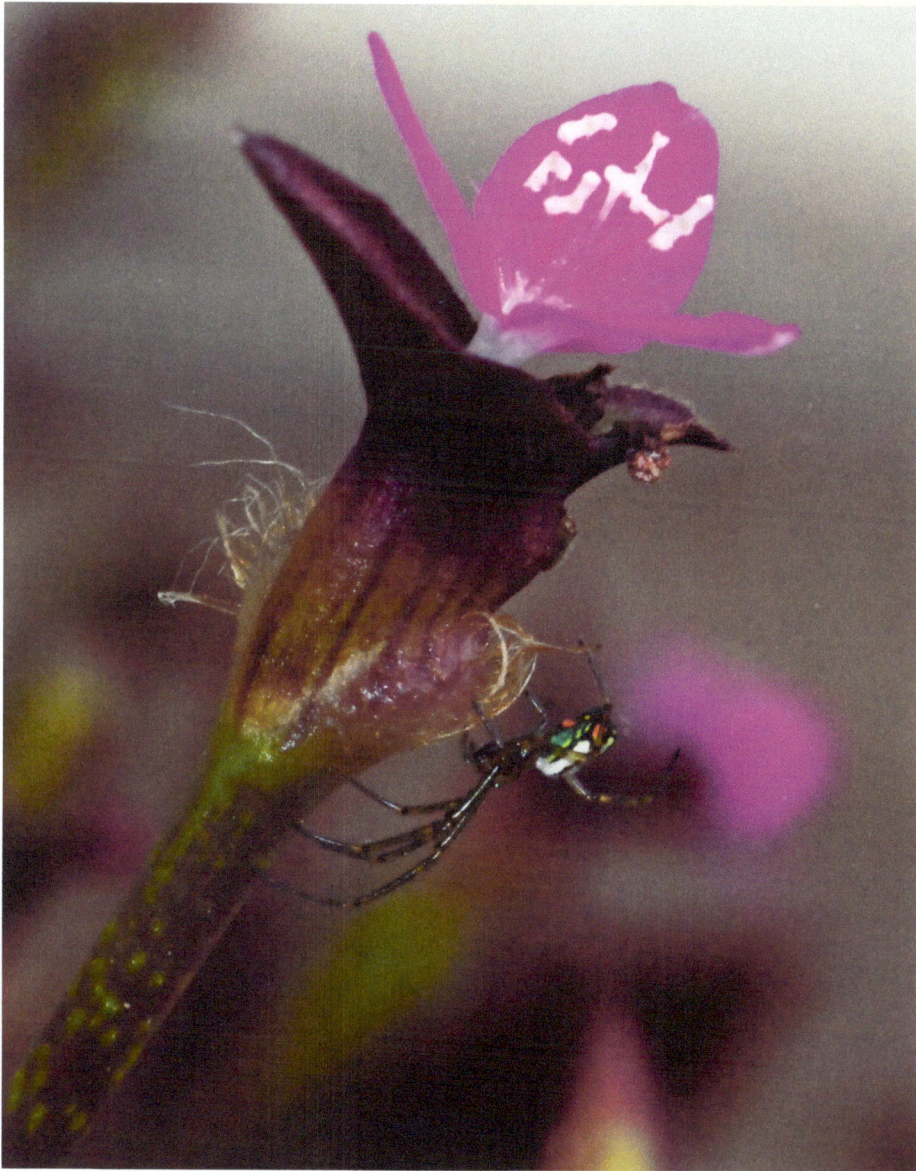

A spider begins the first threads of its web. June 2009. La Hesperia Biological Station, Ecuador.

HANGING BY A THREAD

Flora and Fauna of the Amazon Rainforest and Cloudforest

GLASSWING BUTTERFLY

A rare glasswing butterfly (Greta oto) has the ultimate camouflage of transparent wings. June 2009. La Hesperia Biological Station, Ecuador.

A marsupial tree frog (Gastrotheca sp.) sits triumphant on its tulip. June 2009. Azama, Ecuador.

THE VIEW

TRANSFORMATION

A beetle metamorphosizes within a cocoon of mucous. February 2010. Yanayacu Biological Station, Ecuador.

A hachupalla plant (Puya hamata) growing in the highland Andean paramo ecosystem. It is the staple food for the rare spectacled bear (Tremarctos ornatus). February 2010. La Guandera Biological and Research Station, Ecuador.

SPECTACLED BEAR TRAP

Flora and Fauna of the Amazon Rainforest and Cloudforest

LAVENDER WIND

Flowers blowing in the wind around the highland Andean lake of Laguna de Mojanda. June 2009. Otavalo, Ecuador.

A rarely blooming native cloudforest orchid (Oncidium nubigenum). February 2010. La Guandera Biological and Research Station, Ecuador.

MIST GROWING ORCHID

Flora and Fauna of the Amazon Rainforest and Cloudforest

BROMILIADE

Wild broliades (family Bromiliaceae) commonly grow on tree branches in the Andean cloudforest. February 2010. La Guandera Biological and Research Station, Ecuador.

BAMBOO SHOOT

Contrary to its unique appearance, bamboo is actually a group of perennial evergreen. It is the fastest growing plant in the world, soaring skyward at an almost visible rate. February 2010. La Guandera Biological and Research Station, Ecuador.

BAMBOO INTERIOR

The spiraled interior maze of a bamboo stock holds crystal clear potable water that the plant has filtered from the ground. June 2009. La Hesperia Biological Station, Ecuador.

Flora and Fauna of the Amazon Rainforest and Cloudforest

Within the plethora of water and mist of the cloudforest, ferns abound. La Hesperia Biological Station, Ecuador.

FERN CURL

Flora and Fauna of the Amazon Rainforest and Cloudforest

THE SUN SETS ON
THE AMAZON

AFTERWORD

Lindsay Erin Lough's passions in life were dancing, photography, writing, and caring for her chinchillas. She graduated with highest honors from Princeton University with a degree in Ecology and Evolutionary Biology and a minor (License) from the Classics Department. She was a published research scientist and she had completed three years of medical school at the Mayo Clinic. She held a Master's degree in journalism from the Walter Cronkite School at Arizona State and was thinking of becoming a medical journalist.

In 2011 she published a book of her Amazon nature photography taken while living and traveling in Peru and Ecuador. She was teaching photography, writing an autobiography, and reprising her modern ballet skills at the time of her death in Phoenix, Arizona.

Surprisingly perhaps, given her accomplishments, Erin suffered from various forms of severe mental illness throughout adulthood. She sought treatment continuously with periods of remission, sometimes in conjunction with inpatient care. She fought hard to lead an independent, dignified, and productive life. But her condition would always worsen again despite her access to professional medical providers and the support of family and friends. During her last two years, her deteriorating mental health consumed her life. No one seemed able to help her.

Unfortunately, Erin's experience is not unusual for many people suffering from mental illness. Erin understood the risks and stigma of her situation in society, even as she strove to find a place for herself in that society. While the medical profession has had occasional success treating individuals like Erin, the numbers of people afflicted are large and growing. The need for more research, for better protocols of intervention and care, and for greatly expanding the numbers of caregivers and facilities has become obvious.

Erin worked tirelessly to demonstrate the importance of acceptance and compassion, as well as professionalism and competence, when treating the suffering. She aimed to lead a professional life that would mitigate the negative stereotyping of mental illness and improve current medical practices which are failing so many individuals like her.

Erin died on May 12, 2016, at the age of thirty-one. In remembrance, all royalties earned from the publication of this book of her Amazon nature photography will be donated to organizations that are actively working to improve the care and treatment given to those with mental health disease.

www.ingramcontent.com/pod-product-compliance
Lightning Source LLC
Chambersburg PA
CBHW041542260326
41914CB00015B/1523